AMAZING MILITARY M

AMPHIBIOUS VEHICLES

BY ELEANOR BRADSHAW

Enslow PUBLISHING

Please visit our website, www.enslow.com. For a free color catalog of all our high-quality books, call toll free 1-800-398-2504 or fax 1-877-980-4454.

Library of Congress Cataloging-in-Publication Data
Names: Bradshaw, Eleanor, author.
Title: Amphibious vehicles / Eleanor Bradshaw.
Description: Buffalo, New York : Enslow Publishing, 2025. | Series: Amazing military machines | Includes bibliographical references and index.
Identifiers: LCCN 2024000196 | ISBN 9781978539594 (library binding) | ISBN 9781978539587 (paperback) | ISBN 9781978539600 (ebook)
Subjects: LCSH: Motor vehicles, Amphibious–United States–Juvenile literature. | Amphibious assault ships–United States–Juvenile literature. | Military vehicles–United States–Juvenile literature.
Classification: LCC UG446.5 .B68267 2025 | DDC 629.2/9–dc23/eng/20240203
LC record available at https://lccn.loc.gov/2024000196

Published in 2025 by
Enslow Publishing
2544 Clinton Street
Buffalo, NY 14224

Portions of this work were originally authored by Ryan Nagelhout and published as *Amphibious Vehicles*. All new material in this edition is authored by Eleanor Bradshaw.

Designer: Andrea Davison-Bartolotta
Editor: Rachael Morlock

Photo credits: Cover U.S. Marine Corps photo by Sgt. Trystan Jordan/DVIDS; series art (metal background texture) caesart/Shutterstock.com; series art (title font) TortugaStudios/Shutterstock.com; p. 5 U.S. Navy photo by Mass Communication Specialist 2nd Class Chris Roys/Released/DVIDS; p. 7 (inset) File:PSM V12 D470 Oliver Evans.jpg/Popular Science Monthly/Wikimedia Commons; p. 7 (main) courtesy of Library of Congress; p. 9 humberto valladares/Alamy Stock Photo; p. 10 U.S. Naval History and Heritage Command Photograph/Navy.mil; p. 11 National Museum of the U.S. Navy/Flickr; p. 12 courtesy of the National Archives; p. 13 Gary Todd/Flickr; p. 15 U.S. Navy photo by Mass Communication Specialist 3rd Class Kevin V. Cunningham/Released/DVIDS; p. 17 U.S. Marine Corps photo by Lance Cpl. Brendan Mullin; p. 19 U.S. Marine Corps photo by Ashley Calingo/DVIDS; p. 20 U.S. Marine Corps photo by Lance Cpl. Willow Marshall/DVIDS; p. 21 U.S. Marine Corps photo by Sgt. Courtney G. White/DVIDS.

Printed in the United States of America

CPSIA compliance information: Batch #CS25ENS: For further information contact Enslow Publishing, at 1-800-398-2504.

CONTENTS

Crossing Land and Water 4
Early Attempts 6
Built for War 8
Marine Machines 10
From LVT to AAV 14
The ACV 18
Ready for Action 20
Words to Know 22
For More Information 23
Index 24

Boldface words appear in Words to Know.

CROSSING LAND AND WATER

Can you imagine a boat that drives on land, or a car that floats at sea? A machine that does both is called an amphibious vehicle. The U.S. military uses these vehicles to move people and things across water and land.

Amphibious vehicles **launch** from and return to large ships at sea.

EARLY ATTEMPTS

One of the first amphibious vehicles was made in 1805 by Oliver Evans in Philadelphia, Pennsylvania. Evans called it *Orukter Amphibolos*, Greek for "amphibious digger." The machine ran on steam. Evans made it for digging up the bottoms of rivers.

Evans's *Orukter Amphibolos* was 30 feet (9.1 m) long and 12 feet (3.7 m) wide.

OLIVER EVANS

BUILT FOR WAR

The U.S. military started making amphibious vehicles during **World War II**. One early vehicle was called the DUKW, or "Duck." The army used it to move soldiers and supplies from sea to land. DUKWs had six wheels and a **propeller**.

The U.S. Army made the DUKW in 1942. It looked like a boat on wheels.

MARINE MACHINES

The U.S. Marine Corps used another type of amphibious vehicle in World War II. It was called the Landing Vehicle Tracked (LVT). It used **tracks** instead of wheels to cross the ground. It could move through water and drive on beaches.

The LVT looked more like a tank than a boat.

Two types of LVTs were built. One had **armor** to guard from **attack** as it moved people and **cargo**. The other had armor and a large gun for firing at enemies. After World War II, LVTs were important in the Korean War.

Marines used LVTs in the Korean War, which was fought from 1950 to 1953.

FROM LVT TO AAV

A new machine entered service in 1972. It was called the Amphibious **Assault** Vehicle, or AAV. AAVs were made to be the first vehicles to reach land during beach assaults. They could also fire machine guns and launch **grenades**.

AAVs traveled at 45 miles (72 km) per hour on land and about 8 miles (13 km) per hour in water.

For many years, the U.S. military used the AAV-7. It could hold 25 people and move 10,000 pounds (4,536 kg) of cargo. The hull, or body, was made of **aluminum** to keep marines safe. The vehicle weighed about 29 tons (26 mt)!

A crew of three people—a commander, a gunner, and a driver—kept an AAV running.

THE ACV

The U.S. Marine Corps now uses the Amphibious **Combat** Vehicle, or ACV. New ACVs are being built and tested. Soon, ACVs will replace AAVs in the U.S. military. ACVs are faster on land, safer at sea, and better built for combat.

GUNS
CARGO BAY
HULL
WHEELS

READY FOR ACTION

The first amphibious vehicles in the military moved soldiers from ship to shore. From DUKW to ACV, these machines have changed a great deal. Today's amphibious vehicles are not only boats that can drive, they're also vehicles that can attack.

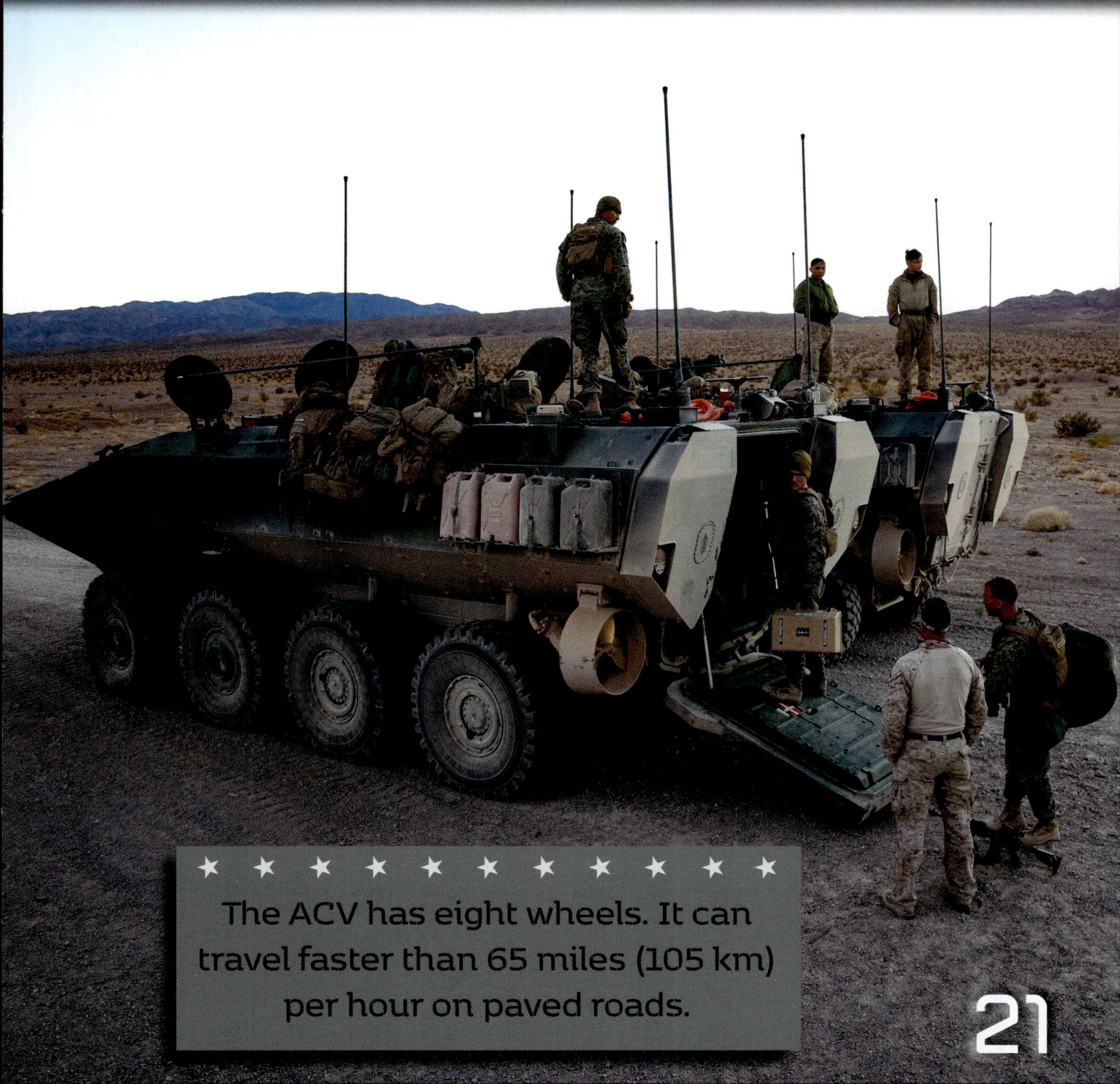

The ACV has eight wheels. It can travel faster than 65 miles (105 km) per hour on paved roads.

WORDS TO KNOW

aluminum: A type of lightweight metal.

armor: A thick covering used to keep someone safe from harm.

assault: A sudden attack.

attack: To try to harm someone or something.

cargo: Supplies or tools being moved.

combat: Armed fighting between opposing forces.

grenade: A small bomb thrown by hand or using a launcher.

launch: To send out with great force.

propeller: Paddle-like parts on a ship that spin in the water to move the ship forward.

track: An endless belt on which a vehicle travels.

World War II: A war fought from 1939 to 1945 that involved countries around the world.

FOR MORE INFORMATION

BOOKS

London, Martha. *US Marine Corps: Equipment and Vehicles*. Minneapolis, Minnesota: Kids Core, 2022.

Sprott, Gary. *Amphibious Vehicles*. Vero Beach, Florida: Rourke Educational Media, 2020.

WEBSITES

Brief History of the United States Marine Corps
www.usmcu.edu/Research/Marine-Corps-History-Division/Brief-Histories/Brief-History-of-the-United-States-Marine-Corps/
Read a short overview of the United States Marine Corps and its role in various wars through history.

World War II
kids.britannica.com/kids/article/World-War-II/353934
Learn more about World War II, the war in which amphibious vehicles were first used, in this encyclopedia article.

INDEX

Amphibious Assault Vehicle (AAV), 14, 15, 16, 17, 18

Amphibious Combat Vehicle (ACV), 18, 20, 21

armor, 12

DUKW, 8, 9, 20

Evans, Oliver, 6, 7

Korean War, 12, 13

Landing Vehicle Tracked (LVT), 10, 11, 12, 13

Orukter Amphibolos, 6, 7

tracks, 10

U.S. Army, 9

U.S. Marine Corps, 10, 13, 16, 18

U.S. military, 4, 8, 18

wheels, 8, 9, 10, 21

World War II, 8, 10, 12